Filipe Heringer S. do Amaral

Sizing the Fire Prevention and Protection Plan

Filipe Heringer S. do Amaral

Sizing the Fire Prevention and Protection Plan

Case study of a residential building

ScienciaScripts

Imprint
Any brand names and product names mentioned in this book are subject to trademark, brand or patent protection and are trademarks or registered trademarks of their respective holders. The use of brand names, product names, common names, trade names, product descriptions etc. even without a particular marking in this work is in no way to be construed to mean that such names may be regarded as unrestricted in respect of trademark and brand protection legislation and could thus be used by anyone.

Cover image: www.ingimage.com

This book is a translation from the original published under ISBN 978-613-9-60482-1.

Publisher:
Sciencia Scripts
is a trademark of
Dodo Books Indian Ocean Ltd. and OmniScriptum S.R.L publishing group

120 High Road, East Finchley, London, N2 9ED, United Kingdom
Str. Armeneasca 28/1, office 1, Chisinau MD-2012, Republic of Moldova, Europe
Printed at: see last page
ISBN: 978-620-7-28389-7

ACKNOWLEDGEMENTS

I would firstly like to thank God for making this dream possible and for always being there for me at every stage of my life.

To Prof. MSc. Pedro Genuino de Santana Junior, for his zealous guidance, invaluable ideas and above all, for his commitment, enthusiasm and dedication at every stage.

To my family, who directly or indirectly contributed to the realisation of this project, with words of encouragement to keep me going and not let obstacles get the better of me. To my parents Cleofas and Analice, role models in my life, for all the education and humility they passed on to me. Especially my mum, my eternal companion and strength, and my wife Joiceane, who was by my side throughout the long journey. To the rest of my siblings, friends and classmates who encouraged me in some way along the way.

This is just one of the many achievements I'm aiming for, and I'll be taking the next steps towards new achievements in my life, because I'm only just starting out.

To FACULDADE VÉRTICE - UNIVÉRTIX, its teaching staff and organisers, who provided the opportunity for this achievement.

To the supervisor Prof. MSc. Pedro Genuino de Santana Junior, for his commitment and support in the preparation of this work.

Mrs Walquíria Pereira, who made it possible to carry out the field research in her building, thus enabling me to write this paper.

Filipe Heringer Silveira do Amaral

SUMMARY

The aim of this study was to verify, identify, analyse and propose a revision of the Fire Prevention and Combat Plan (PPCI) for the Axel Gomes Building, improving the performance of the most critical locations, taking into account the existing infrastructure and compliance with the standards in force during the process for issuing the Inspection Certificate from the Minas Gerais Fire Department - AVCB. This analysis was carried out through a case study, taking into account the legislation in force (decrees, provisional measures, legislation, technical standards and technical instructions from the Minas Gerais Fire Department) and the current context. Based on the perspective of prevention, this work aims to make a contribution by diagnosing and analysing the safety conditions of a multi-family building, proposing some interventions that are necessary to improve the safety of this building, so that we can encourage a rethink of the best strategies employed to hinder the presence of fire risk factors in this residential building.

KEYWORDS: Design, Prevention, Safety, Fire.

SUMMARY

CHAPTER 1

INTRODUCTION

Man has always wanted to master fire. For thousands of years, striking one stone against another would generate a spark which, together with sticks, would start a fire. He controlled the ignition. However, he didn't control the spread of the fire (Seito, 2008).

The eruptions of volcanoes, atmospheric discharges, now known as natural phenomena, in ancient times were associated with the wrath of the gods, a real punishment from heaven. Fire itself was venerated. The Incas, in the Machu Pichu region of Peru, greeted the sunrise with loud kisses, offered with their hands, because they considered their leaders to be descendants of the God Inti (Sun) (Bonitese, 2007).

The mastery of fire enabled great advances in knowledge: cooking food, making ceramic pots and jars or glass objects, forging steel, fireworks, etc. On the other hand, there has always been a loss of life and property due to fires (Bonitese, 2007).

According to Seito (2008), after the Second World War fire began to be seen as a complex science, as it involved knowledge of physics, chemistry, human behaviour, toxicology, engineering, etc.

For decades, human beings have been learning the techniques for mastering this powerful and overwhelming energy source, but in many situations there has not been enough time to implement standards, procedures and systems to enable effective control in buildings, as seen in the countless historic fires that have occurred in Brazil, such as the Joelma Building, with 25 floors, in 1974; the Andraus Building, with 31 floors, in 1972; the CESPI towers, with 21 and 27 floors, in 1987, both in São Paulo; among others. These fires marked the country in the cruelest possible way, with the loss of hundreds of people's lives, important documents, incalculable material damage and a collective phobia of fire in large buildings (Fagundes, 2013).

These events triggered a national concern about fire safety in buildings. It was the awakening to the problem in Brazil, generating various pieces of legislation established by the competent bodies, such as the Fire Brigade, the Public Prosecutor's Office, the Ministry of Labour and Employment, among others (Fagundes, 2013).

According to the Federal Council of Engineering and Agronomy (CONFEA), resolution N^0 1.010, of 22 August 2005, in the context of engineering there is always concern about the stability and durability of works, as well as their perfect functioning. Both stability and comfort in the use of these products must be guaranteed, especially the safety of users. Among the numerous fields of activity of the civil engineer described in Annex II of

Resolution No. 1.010 of the Federal Council of Engineering, Architecture and Agronomy CONFEA is that of: Fire Prevention and Fighting Installations, Equipment, Components and Devices.

According to Seito et al (2008), fire safety is regarded internationally as a science and therefore an area of research, development and teaching. It is an international trend to demand that all materials, components and construction systems, equipment and utensils used in buildings be tested from the point of view of fire safety.

These international developments can be seen in countries such as the United States, Japan and some European countries. In developing countries, this issue has not yet been given due importance (Seito, 2008).

In most of the country's engineering and architecture schools, this topic of fire safety is covered in a tangential way, and does not train professionals with sufficient capacity to deal with the challenges posed by the subject. In many of the country's urban centres, there is negligence on the part of business owners, inspection bodies in general, and even designers and builders, when it comes to inspecting and ensuring the perfect functioning of fire prevention and firefighting installations (Seito, 2008).

On this subject, it's important to note that tragedies and accidents of this nature continue to happen, such as the recent case of the Kiss nightclub fire in January 2013 in Santa Maria/RS. This case suggests a significant lack of safety culture, whether due to negligence on the part of the owners or neglect on the part of public bodies. According to the Technical Report presented by the Regional Engineering Council of Rio Grande do Sul (CREA/RS 2013), the fire occurred inside a shed that had been wrongly adapted to be a bar or discos. The architectural changes made did not take into account the possible risks and the investments made in the area of safety were misdirected or insufficient (Fagundes, 2013).

Fire protection must be seen as an obligation and necessity to protect human lives first and foremost, and secondarily the property involved, regardless of its financial cost. The construction of any type of safer building should be an unavoidable and ethical duty of the designer, builder and entrepreneur, regardless of legal requirements (Corpo de Bombeiros, 2000).

Therefore, the purpose of this work is to provide technical support, according to the process of classifying the building according to the Fire Brigade's Technical Instructions, identifying the types of fire-fighting equipment to be used, sizing the fire-fighting systems and the documentation needed to file the project with the responsible bodies, regarding the multi-family building, in this case the Axel Gomes Building, located in the city of Matipó/MG, at Rua Evêncio de Abreu, N⁰ 34, Centro.

With this study I intend to contribute to engineering professionals, occupational safety technicians, masters and academics in the process of implementing and regularising the Fire Prevention and Combat Plan (PPCI) with the Minas Gerais Military Fire Brigade - CBMMG, a subject that is extremely important for everyone's professional life.

CHAPTER 2

THEORETICAL FRAMEWORK

2.1 - FIRE

Every building, from a safety point of view, is subject to an unpredictable calamity: fire. Fire is capable of causing major accidents and catastrophes, resulting in loss of life and enormous material damage. In order to prevent and fight fires effectively, one must know the mechanics of fire in all its aspects: causes, formation and consequences (Fagundes, 2013).

Modern life increases the risk of fires due to the large concentrations of people in the closest and tallest buildings, architectural designs that favour the spread of fire and materials used that are easy to combust (Seito, 2008).

"Fire will always coexist with man, so both must live in harmony and, for this to happen, it must be controlled so that this relationship is not broken" (Brentano, 2010).

Fire can be defined as a chemical reaction, called combustion, which is a rapid oxidation between a combustible material, solid, liquid or gaseous, and the oxygen in the air, caused by a heat source that generates light and heat (Seito, 2008).

In other words, fire is a living combustion that manifests itself through the production of flames that generate light and give off heat, as well as the emission of smoke, gases and other residues. According to Brentano (2010), each of these combustion products has consequences:

The flames form the spectacular and visible part of the fire, illuminating and attracting, while the smoke impedes visibility, causes panic, intoxication and/or asphyxiation, making it difficult to get out and approach the fire Brentano (2010).

Gases are invisible, can be toxic, odourless and their diffusion causes the fire to spread. Nowadays, with more and more synthetic materials used in building cladding, the amount of gaseous products harmful to humans in a fire situation has increased. Smoke and toxic gases are responsible for more than 80 per cent of deaths in fires Brentano (2010).

Heat warms the air to very high temperatures, causing fire to spread through the spontaneous combustion of certain materials and the deformation and loss of resistance of others, such as the very structure of a building Brentano (2010).

The oxygen in the air is consumed during combustion in closed environments, making it unbreathable Brentano (2010).

The waste left behind by common solid fuels, such as ash, generates environmental pollution Brentano (2010).

For fire to occur, three essential elements must be present simultaneously: combustible material, oxidiser (oxygen) and a heat source, forming the fire triangle. If the fire spreads after it has occurred, there must be a transfer of heat from molecule to molecule of the combustible material, still intact, which combusts successively, generating a chemical chain reaction (Brentano, 2010).

According to Seito (2008), the main characteristics of the component elements of fire are:

Fuel is any material that is susceptible to burning, i.e. after ignition, it continues to burn without any additional addition of heat. It can be solid, liquid or gaseous. Most solid fuels have a sequential ignition mechanism. To ignite, they must first be heated, releasing combustible vapours that mix with the oxygen in the air to create a flammable mixture. Liquid fuels vaporise when heated, mixing with the oxygen in the air to form a flammable mixture. In order to combust, gases must form a flammable mixture with the oxygen in the air, the concentration of which must be within an ideal range Seito (2008).

The oxidising agent, usually oxygen from the air, is the chemical agent that activates and maintains combustion, combining with the fuel gases or vapours to form a flammable mixture Seito (2008).

Heat is the energy that starts, maintains and encourages the spread of fire. Heat causes the chemical reaction of the flammable mixture, coming from the combination of gases or vapours from the fuel and the oxidiser Seito (2008).

The chemical chain reaction, which is the transfer of heat from one molecule of the burning material to the neighbouring molecule, still intact, which heats up and combusts, and so on, until all the material is burning Seito (2008).

2.2 - FIRE CLASSES

According to Luz Neto (2013), fires are classified according to the combustible material and are divided into four classes:

- Class A: These are fires that occur in common combustible materials such as wood, paper, fabrics, etc. These materials burn on the surface and in depth, leaving residues after combustion, such as embers and ashes. They are best extinguished by water, as these materials need to be cooled to extinguish the fire.

- Class B: These are fires that occur when air mixes with the vapours that form on the surfaces of flammable combustible liquids such as oils, petrol, among others, which burn only on the surface, leaving no residue; and flammable gases such as liquefied petroleum gas (LPG), natural gas (NG), hydrogen and others. They can be extinguished by smothering, by breaking the chemical chain reaction or by removing the combustible material. Extinguishing agents can be dry chemical products, vaporising liquids, CO2, nebulised water and chemical foam, which is the best extinguishing agent.

• Class C: These are fires that occur in energised electrical equipment.

A non-conductive extinguishing agent should be used. Dry chemical powders, vaporising liquids and CO2 are used.

• Class D: These are fires that occur in combustible metals, called pyrophoric, such as magnesium, titanium, lithium, aluminium, among others. These metals burn more quickly, react with atmospheric oxygen and reach higher temperatures than other combustible materials. Combat requires special equipment, techniques and extinguishing agents for each type of combustible metal, which form a protective layer isolating the combustible metal from atmospheric air.

2.3 - BRAZILIAN CONTEXT

Historically, we can identify several accidents involving fires in residential buildings, the consequences of which include high human losses, significant financial losses and significant concern for society as a whole.

According to Luz Neto (2013, p 18):

> The country often witnesses loss of life and huge economic damage, especially in the expanding urban environment. Yet society has not produced a fire protection policy.

We should also remember that for Ono (2007, p 99):

> The area of fire safety gained momentum in the country, specifically in the state of São Paulo, in the first half of the 1970s, when two major fires occurred in the city of São Paulo with international repercussions: the Andraus Building and the Joelma Building.

However, it was as a result of these accidents over time that standards and codes emerged with a view to minimising these events. New times demand new alternatives, transformations and proposals that fulfil the main objective of fire safety Almeida (2002).

However, as Ono (2007, p 99) points out:

> There has been little participation from the main players in this scenario: architects and civil engineers, who are responsible for designing building spaces, specifying materials and carrying out work that effectively guarantees the inclusion of fire safety measures.

In this context, we can observe the increase in buildings in large urban centres, which necessarily leads us to rethink certain aspects. One point to highlight is the fire safety of collective residential buildings, which make up the vast majority of buildings in large urban centres (Fagundes, 2013).

As Almeida (2002) states, the relevance of this work lies in the fact that Brazil demonstrably lacks a culture of prevention and concern about risk factors, especially fire risks.

This panorama calls for immediate action and an intense process of reflection that can result in planning and goals that can benefit everyone (Almeida, 2002).

2.4 - CAUSES OF FIRE

For a fire to start in a building, there must be simultaneous and fundamental competition from a heat source, a fuel and a human component. The human component becomes fundamental in this event and can be found through flaws in the design and/or execution of installations, as well as behavioural negligence when occupying the building. These components, combined with the chemical chain reaction and oxygen, guarantee the maintenance of the fire, as well as its growth (Fagundes, 2013).

According to Pozzobon (2007), when studying the causes of a fire, the aim is to find out how, why and where the combustion process started, whether or not it was caused by direct human action.

2.5 - FIRE SPREAD

According to Callister (2013), fire behaves in a complex way and its spread is often unpredictable. The factors that contribute to the spread of fire are related to the transmission of heat, which can occur in three main ways:

•   Conduction or contact, by the flames themselves passing from one floor to another through windows, curtains and other materials, or through a physical medium heated by the fire that conducts the heat to the other, such as walls and ceilings.

•   Convection, i.e. by the circulating gaseous medium, such as the gases and hot air produced by the fire, which rise and come into contact with other materials that are heated until they combust.

•   Radiation, i.e. by means of waves or heat rays generated by a heated body, which radiates heat in all directions through space, similar to light. It is the thermal sensation felt on the skin due to the sun's rays or when approaching a fire.

In a fire, the three forms of fire propagation are usually concurrent, although at any given time one of them predominates over the others (Bonitese, 2007).

The spread of fire must always be thought about and analysed very carefully when drawing up a fire protection plan, thus eliminating the possibility of a chain reaction (Bonitese, 2007).

2.6 - PREVENTION

According to Luz Neto (2013), prevention comprises a set of measures that try to prevent the onset of illnesses or accidents that could cause an undesirable occurrence. This highlights the importance of care as a way of avoiding or minimising the tragic consequences of unexpected incidents such as fire through risk prevention and the elimination of risk and accident factors, using strategies or instruments that make this possible. According to Luz Neto (2013), evacuation is one of the essential aspects of protecting buildings and people.

This is why we consider the study proposed here on fire safety, aimed at protecting human life, to be of great importance.

According to Bonitese (2007, p 3):

> In the study of fire safety, the need to merge regulatory measures and the architectural design process is imminent, in order to maximise the safety factor in buildings in terms of structural and property protection, as well as safeguarding lives, allied to aspects of habitability.

Luz Neto (2013), on the other hand, emphasises that considering the economic losses and the loss of human lives involved in fires has made it possible to increase research and investigation in this area of knowledge in recent decades.

Fire prevention has had to follow the development of the use and utilisation of fire energy. The more this energy is used, the more it is necessary to invest in projects and programmes based on anticipating the effects of fires. When fire energy was only used to heat cave dwellings, fire prevention was carried out by choosing caves near lakes and rivers. We know how much fire has contributed to humanity, because without it we certainly wouldn't have built the great cities. Its use is indispensable in all spheres of activity, especially in areas of work, leisure and especially for the development of physical, chemical and technological research (Seito, 2008).

According to Almeida (2002), there are currently no studies in the national literature on the risks of fire in urban spaces. However, the author points out that in foreign literature we can highlight well-established studies related to fire prevention in buildings.

2.7 - FIRE EXTINGUISHING METHODS

According to Brentano (2010), whenever you want to extinguish a fire, you have to neutralise at least one of its three components or interrupt the chemical chain reaction.

Fire extinguishing methods are used according to the component element of the fire that is to be neutralised:

- Extinguishing by isolation: In some fire situations it is possible to remove combustible material. In building fires, neutralising this element is difficult, if not impossible.
- Extinguishing by smothering: In this case, the aim is to prevent the burning material from being fed more oxygen from the air, reducing its concentration in the flammable mixture.
- Extinguishing by cooling: With the use of an extinguishing agent, this agent absorbs the heat from the fire and the burning material, consequently cooling the material. In general, cooling the combustible material is the most common way of extinguishing fire in buildings and the most commonly used agent is water.
- Chemical extinguishing: When certain extinguishing agents are thrown on the fire, their molecules dissociate due to the action of heat, forming free radical atoms, which combine with the flammable mixture resulting from the gas or vapour of the combustible material with the oxidising agent, forming a non-flammable mixture, interrupting the chemical chain reaction.

2.8 - EXTINGUISHING AGENTS

According to Callister (2013), each combustible material has its own combustion characteristics, thus requiring specific ways of extinguishing the fire. The extinguishing agent to be used must be appropriate, so that it acts quickly and efficiently, causing minimal damage to people's lives, the contents and the building.

The main extinguishing agents used are: water, aqueous or mechanical foam, inert gases and dry chemical powders (Callister, 2013).

2.9 - FIRE PROTECTION MEASURES

According to Almeida (2002), in order to achieve a degree of efficiency against fire, in terms of design and operation, protection measures are recommended by the technical standards and legislation in force.

Protective measures can be divided into:

- Passive or preventive: These measures aim to minimise the chances of a fire starting and reduce the likelihood of it spreading (Almeida, 2002).

- Active or combat measures: These measures aim to act on the existing fire in order to extinguish it or control it until the fire brigade arrives on the scene, creating facilities for this fight to be as effective as possible (Almeida, 2002).

2.10 -PROJECTS

According to Brentano (2010), human life is the main objective and, as such, should always be thought of as the most important and delineating factor of all the parameters that determine building design.

It is also worth highlighting the protection of assets, as the investments are quite high, and consequently the losses due to a fire (Brentano, 2010).

Many safety measures that must be taken can only be carried out when they are provided for in the architectural project, because they involve the areas and volumes of the buildings. To this end, we can affirm the importance of architectural design as the beginning of fire protection (Almeida, 2002).

The fire prevention project is all the fire protection measures that must be taken in a building, both passive and active, and it must be submitted to the competent public bodies for analysis and approval. The project is made up of a set of written and graphic documents (Almeida, 2002).

According to Seito (2008), the architectural project and the PPCI project must focus on two basic premises:

- Prevent a fire from starting. To this end, the building project must include all the construction measures needed to prevent a fire from occurring.

- In the event of an outbreak of fire, appropriate means must be provided for the safe and rapid evacuation of the building and adequate facilities so that it can be isolated at its place of origin and fought quickly and effectively.

2.11 - CLASSIFICATION OF BUILDINGS

According to the CBMMG's Technical Instruction N⁰ 01, the classifications are extremely important because they are used to define the building's fire prevention conditions and the fire-fighting systems to be used.

2.11.1 - Risk classification of buildings

To draw up the PPCI, it is necessary to check with CBMMG Technical Instruction N^0 09 to define the fire load of the building and consequently the risk level classification: Class A (low risk), Class B (medium risk) and Class C (high risk).

According to NBR 12693/2013, each building's occupancy type is classified according to its specific fire load, which can be low risk, medium risk or high risk.

2.11.2 - Classification of the building according to its occupation

This classification is fundamental for the development of the PPCI, the value of which will be used to determine the population calculation, and its results are important for verifying the protection required when implementing it. For this classification we use Table 1 of NBR 9077/2001.

2.11.3 - Building height classification

According to Brentano (2010), fire protection measures should be implemented at three levels:

- Descending height (hd): This height is defined as the difference in level between the floor of the last type floor or habitable floor and the level of the unloading floor that gives access to the public pavement.
- Rising height (ha): This height is defined as the difference in level between the lowest floor of the building, in this case the basement or the last basement when there is more than one, and the level of the unloading floor giving access to the public pavement.
- Actual or total height (ht): This height is defined as the gradient between the outlet to the public highway of the lowest discharge level and the highest level of any building, usually the top of the upper cold water tank. It is used in the sizing of the lightning protection system (SPDA).

The height of the building is one of the key factors in correctly dimensioning the PPCI, because with buildings getting taller and taller, we need to check the appropriate systems for each situation. For this classification we use Table 2 of NBR 9077/2001.

2.11.4 - Classification of the building in terms of its area or dimensions on plan

According to CBMMG Technical Instruction N^0 01, buildings are classified into two large groups for all occupations with an area:

- Less than or equal to 750 m^2 .

- More than 750 m2.

Buildings are classified according to their dimensions in plan according to Table 3 of NBR 9077/2001.

2.11.5 - Classification of the building according to its construction characteristics

Buildings are classified according to their construction characteristics in accordance with Table 4 of NBR 9077/2001.

2.11.6 - Classification of the building according to its fire load

To understand this classification, one must understand the designation of fire load and specific fire load (Fagundes, 2013).

The fire load in a building is the sum of the heat energies that can be released by the complete combustion of all combustible materials contained in a room, floor or building, including wall coverings, partitions, floors and ceilings (Seito, 2008).

Specific fire load is the value of the total fire load divided by the corresponding floor area, expressed in Megajoules per square metre (Mj/m2) (Seito, 2008).

Specific fire loads can be determined by characteristic values in buildings and risk areas, according to occupancy and specific use, in accordance with Table A.1 of NBR 12693/2013.

According to NBR 12693/2013, buildings can be classified as low risk, medium risk and high risk in terms of their specific fire load.

According to Brentano (2010), buildings can be classified according to their specific fire load as detailed in Table 1 of NBR 9077/2001.

2.12 - DETAILING FIRE PROTECTION MEASURES

2.12.1 - Population calculation

The importance of calculating the population of a building is due to the fact that it provides data for sizing emergency exits, regardless of the actual number of occupants (Seito, 2008).

According to Brentano (2010), the population calculation is determined by its occupancy, floor or building area, obtained from the architectural project and its occupational density, obtained from Table 4 of NBR 9077/2001.

The formula used is:

$$P = A x \, Do$$

Where: P = Population in number of people,

A: Area of the room, floor or building in m^2 , Do: Occupational density, in n^0 people/m2.

2.12.2 - Emergency exits

According to Technique No 08 of the CBMMG, the emergency exit or emergency exit route is a continuous, properly protected, signposted and illuminated path, consisting of doors, corridors, vestibules, stairs, ramps, lobbies, external passages, etc., to be travelled by the occupants, by their own means, in the event of a fire or other emergency, from any point in the building, until they reach the public highway or another suitably safe external space.

According to NBR 9077/2001, the basic objectives of emergency exits are to enable occupants to move safely by their own means from any point in the building to a place free from the action of fire, heat, smoke and gases, regardless of the origin of the fire. It must also allow external access for the fire brigade to quickly and safely rescue the occupants.

Emergency exits must meet the legal requirements for building accessibility set out in NBR 9050/2015. They must also have an accessible route for a continuous, unobstructed and signposted path that connects building environments and can be used autonomously and safely by all people, including those with physical disabilities or reduced mobility.

Emergency exit routes generally comprise:

•	On the horizontal plane: These are all the paths or spaces located inside the floors, which can give access to a refuge area on the same floor or directly to the stairs, ramps or emergency lifts. These routes can be corridors, walkways, balconies, terraces, balconies, etc.

•	On the vertical plane: These are all the routes or means used to move between floors of different levels, which give access to refuge areas or the unloading floor. These routes are stairs, ramps and emergency lifts.

The calculation elements required for dimensioning a building's emergency exit routes are:

- Calculating the population according to occupation.
- Calculation of the number of passage units required.
- Maximum distances to be travelled.
- Determining the minimum number of emergency exits.
- Time needed to completely vacate the building.

2.12.3 - Calculating the number of passage units

According to NBR 9077/2001, the width of exits must be sized according to the population that will pass through them. Accesses are sized according to the population of each floor, but stairs, ramps and discharges are sized according to the floor with the largest population, taking into account the direction of the exit.

To calculate the number of passage units required on emergency exit routes, the formula is used:

$$N = P/C$$

Where: N = Number of passage units,

P = Population of the room, floor or building, in n^0 people,

C = Capacity of the passage unit, in number of people per minute/passage unit, according to the occupancy of the building, in accordance with table 5 of NBR 9077/2001.

The minimum widths of emergency exits stipulated by NBR 9077/2001 are as follows:

a)	1.10 m, corresponding to two passage units and 55 cm, for occupancies in general, subject to the following provisions;

b)	2.20 m, to allow the passage of stretchers, beds, and others, in occupancies in group H, division H-3.

The widths of exits must be measured at their narrowest part, and protruding sides, pillars and others larger than 10x25 cm are not permitted (Law 14.130 of 19 December 2001).

2.12.4 - Maximum distances to be travelled

It consists of the distance between the furthest point and access to a safe emergency exit, and must always take into account the risk to human life resulting from the fire. It can vary according to occupancy, the building's construction characteristics and the existence of automatic showers to contain fires (Seito, 2008).

The maximum distances to be travelled can be found in Table 6 of NBR9077/2001.

According to Brentano (2010), exits and staircases must be located in such a way as to effectively give occupants the opportunity to choose the best exit route, but to do so they must be sufficiently distant from each other.

2.12.5 - Discharge

Unloading or unloading area is the section of a building's emergency exit route consisting of the space between the end of a staircase, ramp or emergency lift and a door, which gives access to a protected external area or to the public highway (Law N^0 14.130, of 19 December 2001).

The discharge may consist of:

* Enclosed corridor or lobby,

* Open area with pilotis,

* Open-air corridor.

2.12.6 - Time needed to vacate

An important factor in emergency exits is the time needed for the entire building to be vacated in a fire situation, taking into account the diversity of possibilities for movement and the speed at which occupants move (Brentano, 2010).

According to Brentano (2010), the following is recommended for average commuting speeds and maximum unoccupied time:

- Travel speed:

* Horizontal journeys = 20 m/min.

* Stairs = 5 m/min.

- Maximum time to completely vacate a building = 20 min.

2.12.7 - Corridors

Corridors play a fundamental role in the evacuation of a building, and it is recommended that they have walls lined with fire-resistant materials that do not release toxic gases and smoke. In addition, long corridors must have exhaust openings (Law N^0 14.130, of 19 December 2001).

Runners must arbitrate the following considerations:

- Allow easy flow for all occupants of the building's floors.
- Remain completely unobstructed and free of any obstacles on all floors.
- Have minimum widths according to the passage units.
- They must have a minimum ceiling height of 2.50 metres, in accordance with NBR 9077/2001.
- Visual, audible and tactile signage near the doors leading to the stairs and on the handrails.
- With maximum gradients of 5 mm. In the form of a ramp with gradients of 5 to 15 mm. In the form of steps and signposted as such, above 15 mm.

2.12.8 - Stairs

According to Law No. 14.130 of 19 December 2001, staircases can have different shapes, widths and treads, but any staircase in a building must be non-combustible, the structural elements must offer fire resistance of at least 2 h, they must have guardrails on their open sides and handrails, and they must have non-slip floors and landings.

To correctly size a staircase, the width of the staircase, the height and width of the steps, the flange and the length of the landings must be taken into account.

According to NBR 9077/2001, steps must fulfil the following geometric conditions:

a) The height of the mirror "h" should be between 16 and 18 cm, with a tolerance of 0.5 cm.

b) Have the width of the floor base "b" calculated by the formula:

$$63 \text{ cm} \leq (2h + b) \leq 64 \text{ cm}$$

Stairs must have a landing at least every 3.70 metres and whenever there is a change in direction.

Stair landings must fulfil the following conditions:

- They can't have steps.
- When changing direction, they must be completely flat, with a minimum width equal to the width of the staircase they serve.
- Their length must be calculated according to Blondel's formula, regardless of the width of the staircase;

$$P = (2h + b)n + b$$

Where: P = Length of landing in cm, h = Height of step mirror in cm, b = Width of step base in cm, n = Whole number equal to 1, 2 or 3.

2.12.9 - Guardrails

According to Law N⁰ 14.130, of 19 December 2001, stairs and ramps at emergency exits must be protected on both sides by continuous walls or guardrails whenever there is any unevenness greater than 19 cm.

The height of railings internally must be at least 1.05 metres along landings, corridors, mezzanines and so on; it can be reduced to 92 cm on internal staircases, when measured vertically from the top of the railings to a line at the tips of the jambs or corners of the steps.

The height of railings on external staircases, their landings, balconies and the like, when more than 12 metres above the adjacent ground, must be at least 1.30 metres.

Guards made up of balustrades, railings, screens and the like must have openings of no more than 15 cm in diameter.

2.12.10 - Handrails

Handrails are fundamental elements in the organisation and transit of the population in an emergency exit during a fire situation, reducing the possibility of falls and the consequent obstructions in the way (Brentano, 2010).

In addition, handrails must have a rounded shape that is easily adapted to the anatomical shape of the hand, allowing them to be grasped easily, and must have a continuous, easy and comfortable glide along their entire length. Handrails made of sharp-edged elements are not acceptable for emergency exits (Brentano, 2010).

According to NBR 9077/2001, handrails must be at least 40 mm away from the walls or railings to which they are attached, and extend at least 30 cm beyond the projection of the first step. They must also have a circular or semicircular section with a diameter of between 38 and 65 mm, and their height must be between 80 and 92 cm above floor level. The ends of intermediate handrails must be fitted with balusters or other devices to prevent accidents.

2.12.11 - Emergency signalling

The emergency signalling project must be drawn up using the procedures set out in the technical standards: NBR 9077/2011 - Emergency exits in buildings, NBR 13434-1/2004 - Fire and panic safety signalling - Part 1: Design principles, NBR 13434-2/2004 - Fire and panic safety signalling - Part 2:

Symbols and their shapes, dimensions and colours and NBR 13434-3/2005 - Fire and panic safety signs - Part 3: Requirements and test methods.

According to NBR 9077/2001, exit signs are compulsory for accesses and discharges from emergency staircases in general, in non-residential buildings (i.e. excluding buildings in group A); in places where the public gathers (group F), even if they don't have staircases; and in buildings in occupations B, C, D, E, H, when classified as O (area greater than 750 m).[2]

The purpose of emergency signalling is to orientate and guide the occupants of a building, and it can be both preventative and active in the fight against fire. They aim to identify and warn of potential fire hazards in order to reduce the occurrence of fires, guide the location of fire-fighting equipment and indicate emergency exits, among other things (Fagundes, 2013).

Emergency signalling has four distinct categories, according to their function, which we refer to as signalling conditions for guidance and rescue, warning, prohibition and indication of fire-fighting equipment (Fagundes, 2013).

According to Brentano (2010), emergency signs are differentiated by the colour of each nameplate within a fire prevention and protection system:

- Red: Identifies prohibition signs or identifies fire-fighting and alarm equipment.

- Yellow: Identifies warning signs and danger signals.

- Green: Identifies guidance and rescue signs. The images or symbols on the signposts can be:

- Black: Used on prohibition and warning signs.

- Green: Used on guidance and rescue signs.

- White: Used on equipment nameplates

fire-fighting and guidance and rescue equipment. This should be photoluminescent.

Signs must be fixed to walls at least 1.50 m from the finished floor to the base of the sign, and the maximum distance between them must be between 13 and 15 m (Law[0] 14.130, of 19 December 2001).

Signs on emergency exit route doors must be fixed immediately above the doors, no more than 10 cm above the lintel at the base of the sign; directly on the door leaf, centred, at a height of 1.80 m from the floor to the base of the sign or just above the panic bar with guidance on how to activate it (Law⁰ 14.130, of 19 December 2001).

2.12.12 - Emergency lighting

According to CBMMG Technical Instruction No. 13, the purpose of emergency lighting is to replace normal artificial lighting, which must be switched off or may fail in the event of a fire, with its own energy source that ensures a minimum operating time. During this time, it must guarantee a minimum level of illumination to enable the occupants of a building to leave quickly and safely.

According to NBR 10898/2013, two methods of emergency lighting are possible:

•   Permanent lighting: The emergency lighting lamps are powered by the utility grid and are automatically switched to the alternative power supply in the event of a shortage and/or failure of the normal power supply.

•   Non-permanent lighting: Emergency lighting lamps are not powered by the utility grid and, only in the event of a shortage of this normal source, are automatically powered by the alternative energy source.

There are two types of fire safety lighting:

•   Illumination: This is intended to illuminate exit routes in such a way that occupants have no difficulty evacuating the building.

•   Beacon lighting: This is intended to illuminate obstacles and signposts, and to indicate exit routes, guiding the direction and direction to be taken by the occupants of the building in the event of an emergency.

2.12.13 - Fire extinguishers

NBR 12693/2013 establishes the requirements for the design, selection and installation of portable and wheeled fire extinguishers in buildings and risk areas to combat fire outbreaks.

Fire extinguishers are used as a first line of attack against fires of limited size. They are necessary even if the premises are equipped with automatic showers, hydrants and hoses (Seito, 2008).

This system is compulsory in all buildings except single-family homes, regardless of any other protection measures (Seito, 2008).

There are two types of fire extinguishers: portable ones and those on wheels (carts)

(Seito, 2008).

According to Brentano (2010), the minimum number of fire extinguishers required for fire protection in a building is determined:

- The risk class of the building to be protected and its area,
- The class of fire to be extinguished,
- The selection of the extinguishing agent;
- The extinguishing capacity of the fire extinguisher,
- The maximum area to be protected by a fire extinguisher and the maximum distance to be travelled by the operator,
- The minimum number of fire extinguishers required.

According to NBR 12693/2013, each floor must have at least two extinguishing units, one for class A fire and the other for class B and class C fire. The installation of two ABC powder extinguishing units is permitted. There must also be at least one fire extinguisher no more than 5 metres from the access door to the main entrance of the building, the entrance to the floor or the entrance to the risk area.

Portable fire extinguishers must be installed under the following conditions:

a)   Its handle should be a maximum of 1.60 metres from the finished floor.

b)   The bottom must be at least 0.10 metres from the floor.

finished, even when supported.

It is recommended that fire extinguishers are located where there is the least chance of the fire blocking access, that they are visible and easy to remove, that they remain protected from the weather and potential physical damage and that they are not installed on stairs.

The locations for fire extinguishers must comply with the visual, vertical and horizontal fields, by means of wall signage or painting under the floor in industrial areas and warehouses.

The fire protection system, using extinguishers, must be installed in accordance with the project and undergo maintenance and inspections in accordance with current legislation.

CHAPTER 3

METHODOLOGY

This is a critical analysis of the Fire Prevention and Protection Plan for the building in question, in order to guide effective procedures and methods for meeting the requirements and technical standards in force for drawing up the Fire Prevention and Protection Plan.

The study carried out through quantitative and qualitative evaluation, according to KAUARK, MANHÃES and MEDEIROS (2010) "considers what can be quantified, which means translating opinions and information into numbers in order to classify and analyse them [...]". In accordance with the CBMMG's Technical Instructions, which require a detailed survey of the building, defining its characterisation:

- Built-up area;
- Height of the building;
- Fire load of the building;
- Classification according to Technical Instruction N⁰ 01;
- CBMMG technical assistance form;
- CBMMG inspection report;
- CBMMG fire and panic safety technical project form;
- Technical Project.

3.1 - Description of the object of study

The object of the case study is located at Praça do Café, number 34, in the centre of the city of Matipó - MG. The residential area consists of a 4-storey building with a total height of 18.30 m and a car park on the ground floor, with a total area of 453.13 m² , which is used as a residential condominium.

0 building was built in hollow brick masonry, reinforced concrete structure, waterproofed slab, ceramic floor paving, smooth plaster wall finish with subsequent acrylic paint, ceramic cladding, internal wooden doors, external tempered glass flat doors with aluminium frame, aluminium windows without grilles, external tempered glass staircase door and metal handrails and masonry guardrails. 0 car park follows the same characteristics of materials present in the building as a whole, consisting of 03 metal gates. Figures 1 and 2 show the current building.

Photo 01: front view of the building **Photo 02:** side view of the building Source: Own Authorship

CHAPTER 4

RESULTS AND DISCUSSIONS

This building does not meet the minimum requirements for an AVCB from the CBMMG and consequently does not have a PPCI.

This situation occurs in many urban centres across the country, i.e. negligence on the part of business owners, supervisory bodies in general, and even designers and builders, with regard to the supervision and perfect functioning of fire prevention and firefighting installations (Seito, 2008).Below we will define what should make up a building and what is currently the case:

4.1 - Risk classification of buildings

The building under analysis is classified as a Simplified Technical Project (PTS) in yellow, i.e. low fire risk, according to the classification in Technical Instruction N^0 01 of the CBMMG, as it is a multi-family residential occupation with an area of less than 750 m^2 .

The building is classified as low risk, according to CBMMG Technical Instruction No 01, and its occupancy is residential, described as flats and with a specific fire load of 300 MJ/m2.

4.2 - Classification of the building according to its occupation

In terms of occupancy, according to Table 1 of NBR 9077/2001, the building is classified as a flat block in general, in this case 4 storeys, in group A, division A-2 - multi-family dwellings.

4.3 - Building height classification

The building has a descending height of 10.70 metres. According to Table 2 of NBR 9077/2001, the building is classified with code M - medium height buildings: 6.00 <H < 12.00 m.

4.4 - Classification of the building in terms of its plan dimensions

The building consists of a block of flats and has a private area of 89.50 metres per floor .^{2}

According to Table 3 of NBR 9077/2001, the building is classified with code P - small storey - Sp< 750 itf.

4.5 - Classification of the building according to its construction characteristics

According to Table 4 of NBR 9077/2001, the building is classified with code Z, being the type of building in which the spread of fire is difficult and specifying buildings with a fire-resistant structure and insulation between floors, thus following the recommendation of ideal project types, present in the aforementioned NBR.

4.6 - Population calculation

According to Table 5 of NBR 9077/2001, the building is defined as group A, division A -2, with a population of two people per bedroom.

The 4-storey building has 1 flat per floor, with 2 bedrooms in each flat.

- Total population of the building = 16 people.

4.7 - Risk isolation

According to NBR 9077/2001, the building under study complies with the risk isolation parameters and has horizontal compartmentalisation and vertical compartmentalisation, making it a residential building.

4.8 - Emergency exits and calculating the number of passage units

The escape route project includes the corridors of the floors, the staircase and the unloading.

According to Table 7 of NBR 9077/2001, the number of emergency exits in the building is obtained from the occupancy as group A, division A-2, the dimensions with code P, floor area < 750 m^2 and the height with code M; resulting in 1 N$_{os}$ - a mandatory minimum exit and does not need to be enclosed.

4.9 - Maximum distances to be travelled

According to Table 6 of NBR 9077/2001, a building classified as type Z, occupancy group and division A, without automatic showers and with a single exit can have a maximum distance of up to 40 metres.

4.10 - Discharge

In the building under study, the escape route is discharged at the building's only exit and communicates with its external area.

4.11 - Time needed to vacate

The maximum time to completely vacate this building in a fire situation would be less than 20 minutes, due to the low occupational density, the speed at which its occupants move and the maximum distance to be travelled.

4.12 - Corridors

In this building, the corridors function as a hall per floor, giving access to the flat entrance and staircases.

These corridors have a width of 1.30 metres, a ceiling height of 2.75 metres, no unevenness and only one obstruction on the first floor, as shown in figure 03, thus partially meeting the requirements of NBR 9077/2001.

Photo 03: Obstructed stairwell.
Source: Own authorship

4.13 - Stairs, guardrails and handrails

According to Table 7 of NBR 9077/2001, the type of staircase in the building is obtained from the occupancy as group A, division A-2, the dimensions with code P, floor area $\leq 750m^2$ and the height with code M; resulting in 1 NE - a non-enclosed staircase (common staircase).

This building has a common staircase, in accordance with Blondel's formula (item 2.12.14) with landings 1.30 metres wide and steps with a 0.17 metre mirror and a 0.31 metre base, with a masonry guardrail on one side and a tubular handrail on the other side fixed to the wall, as shown in figures 04 and 05. It therefore partially meets the requirements of NBR 9077/2001, as the handrails are not continuous and should be installed on both sides of the staircase.

Photo 04: Guardrail on the left and **Photo 05:** Discontinuity of the handrail on the right.
Source: Own authorship

4.14 - Emergency signalling

There are no orientation signs throughout the building, as required by NBR 9077/2001, and no fire-fighting signs (fire extinguishers) in accordance with NBR 13434-1/2004.

4.15 - Emergency lighting

The building has an emergency lighting system, one on each stairwell floor, but the luminaires are installed at a height of 2.30m, which is not in accordance with the standard, which requires them to be at a height of 2.50m for voltages of 127 volts, and they do not have INMETRO certification in accordance with the requirements of State Decree No. 44.270 of 1 April 2006 and NBR 10898/2000.

The illumination system is of the lighting type and the method is permanent illumination.

The luminaires total 04 units, 16w fluorescent type, totalling 64W of power, as shown in figure 06.

Photo 06: Emergency light.
Source: Own authorship

4.16 - Fire extinguishers

The building in question is categorised as fire class A, B and C, and as low risk according to its specific fire load.

According to NBR 12.693/2013, each floor must have at least two extinguishing units, one for class A fire and one for class B and C fire, or an ABC powder extinguishing unit, provided it meets the maximum distance to be travelled and capacity. ABC powder extinguishers are mandatory in motor vehicle garages.

The building does not meet the requirements of this standard, as it has no fire extinguishing unit

4.17 - Cost forecasting and technical adjustments

Cost estimates for preventive measures and the implementation of a building's PPCI must be included in the budget spreadsheet from the start of the project.

The costs involved in implementing the PPCI in the building in question totalled R$2,939.97 (two thousand, nine hundred and thirty-nine reais and ninety-seven cents), as shown in Table 1.

Table 1: Cost of implementing the PPCI.

Item	Quantity	Unit Value	Total value

Simplified Technical Project	01	R$ 2.000,00	R$ 2.000,00
ART fee	01	R$ 74,37	R$ 74,37
Public safety fee	01	R$115,60	R$115,60
Signs (exit)	05	R$ 15,00	R$ 75,00
4kg Fire Extinguisher - Class A,B,C	05	R$ 120,00	R$ 600,00
Fire extinguisher signs	05	R$ 15,00	R$ 75,00
Handrail installation	135	R$ 45,00	R$ 6.075,00
	TOTAL		R$ 9.014,97

Source: Author

CHAPTER 5

FINAL CONSIDERATIONS

The aim of this work was to critically analyse the PPCI of a multi-family residential building located in the municipality of Matipó / MG, with regard to in-depth knowledge and technical support for drawing up fire protection projects, emphasising the importance of prevention and planning.

The design of the building in question was analysed from the point of view of: the type of occupancy, the fire class, the construction characteristics, its dimensions, among other issues, in order to assess the fire protection and firefighting system design adopted.

This study showed that the fire prevention and firefighting system should be analysed in conjunction with all the stages of the building's design, being drawn up concurrently with the other projects, because protection is not something that can be added after the building project has been drawn up and this addition compromises its efficiency.

In this context, unfortunately there is still paradox and contempt for the issue of fire prevention in buildings, due to the mistaken idea that the values are taken as costs in the budgeting of the work rather than investment in order to achieve safety for human lives and property.

There is also a general disregard for/unawareness of the rules on the part of building users, who demand that owners add safety devices and training for fire and, consequently, panic situations. It should be emphasised that the time it takes to put the fire safety system into operation is one of the most fundamental factors in controlling or extinguishing the start of a fire, so training people in the equipment provides practical, effective and safe actions.

Therefore, by surveying the costs of implementing the PPCI in the building, we arrive at a derisory figure, i.e. relatively low in relation to the total cost of the work/valuation of the property.

Finally, after studying the prevention and protection methods adopted, it was concluded that the PPCI project prepared complies with the technical regulatory standards of ABNT and the legal requirements of the state of Minas Gerais.

CHAPTER 6

BIBLIOGRAPHICAL REFERENCES

ALMEIDA Junior, Isaac. Analysis of Fire Risks in Revitalised Urban Spaces: An approach in the Recife neighbourhood. Dissertation (Master's in Production Engineering) UFPE, 2002.

ABNT - Brazilian Association of Technical Standards. NBR 10898: Emergency lighting system. Rio de Janeiro, 2013.

ABNT - Brazilian Association of Technical Standards. NBR 11785: Panic bar. Rio de Janeiro, 1997.

ABNT - Brazilian Association of Technical Standards. NBR 12693: Fire extinguisher protection systems. Rio de Janeiro, 2013.

ABNT - Brazilian Association of Technical Standards. NBR 13434-1: Fire and panic safety signalling. Design principles. Rio de Janeiro, 2004.

ABNT - Brazilian Association of Technical Standards. NBR 13434-2: Fire and panic safety signalling. Symbols and their shapes, dimensions and colours. Rio de Janeiro, 2004.

ABNT - Brazilian Association of Technical Standards. NBR 13434-3: Fire and panic safety signalling. Requirements and test methods. Rio de Janeiro, 2005.

ABNT - Brazilian Association of Technical Standards. NBR 9050: Accessibility to buildings, furniture, spaces and urban equipment. Rio de Janeiro, 2015.

ABNT - Brazilian Association of Technical Standards. NBR 9077: Emergency Exits in Buildings. Rio de Janeiro, 2001.

BONITESE, Karina Venâcio. Fire Safety in a Low-Cost Housing Building Structured in Steel. 2007. Dissertation (Master's Degree in Civil Construction) - Federal University of Minas Gerais 2007.

BRAZIL, Fire Brigade. Fire Technology and Manoeuvrability Manual - Digital Instruction

Support System - SIDAI; version 1.0. Rio de Janeiro, 2000.

BRENTANO, T. A proteção contra incêndio ao projeto de edificações. 2º ed. Porto Alegre: T Edições, 2010.

CALLISTER, William D. et al. Materials science and engineering: an introduction. 8.ed. Rio de Janeiro: LTC, 2013.

CONFEA. Resolution no. 1010 of 22 August 2005. Discriminates activities of the different professional modalities of Engineering, Architecture and Agronomy. Available at: www.ufif.br/proengprod/files/2010/04/Resol-Confea- 1010-05.pdf. Accessed on 12 April 2016.

FAGUNDES, Fábio. Fire Prevention and Fighting Plan: Case study in a multi-storey residential building. 2013. Dissertation (Postgraduate Diploma in Occupational Safety Engineering) - Regional University of Northwestern Rio Grande do Sul. 2013.

Fernanda Castro Manhães and Carlos Henrique Medeiros. - Itabuna : Via Litterarum, 2010.

Kauark, Fabiana. Research methodology : a practical guide / Fabiana Kauark, L. A., Falcão Bauer. Construction materials. 5.ed. Rio de Janeiro: LTC, 2014.

LUZ NETO, Manoel Altivo da. Fire safety conditions. Brasília: Ministry of Health, 1995.
MACINTYRE, Archibald Joseph. et al. Plumbing installations: building and industrial. 4.ed. Rio de Janeiro: LTC, 2013.

Minas Gerais, Law Nº 14.130, of 19 December 2001, Technical Instruction - 01 - Technical Procedures of the Military Fire Brigade of Minas Gerais, Military Fire Brigade of Minas Gerais.

Minas Gerais, Law No. 14.130, of 19 December 2001, Technical Instruction - 02 - Fire and Panic Protection Terminology, Military Fire Brigade of Minas Gerais.

Minas Gerais, Law No. 14.130, of 19 December 2001, Technical Instruction - 03 - Graphic Symbols for Fire Safety Projects, Military Fire Brigade of Minas Gerais.

Minas Gerais, Law No. 14.130, of 19 December 2001, Technical Instruction - 04 - Vehicle Access in Buildings and Risk Areas, Military Fire Brigade of Minas Gerais.

Minas Gerais, Law N^0 14.130, of 19 December 2001, Technical Instruction - 06 - Structural Safety of Buildings, Military Fire Brigade of Minas Gerais.

Minas Gerais, Law No. 14.130, of 19 December 2001, Technical Instruction - 08 - Emergency Exits in Buildings, Military Fire Brigade of Minas Gerais.

Minas Gerais, Law No. 14.130, of 19 December 2001, Technical Instruction - 09 - Fire Load in Buildings and Risk Areas, Military Fire Brigade of Minas Gerais.

Minas Gerais, Law No. 14.130, of 19 December 2001, Technical Instruction - 11 - Fire Intervention Plan, Military Fire Brigade of Minas Gerais.
Minas Gerais, Law No. 14.130, of 19 December 2001, Technical Instruction - 12 - Fire Brigade, Military Fire Brigade of Minas Gerais.

Minas Gerais, Law No. 14.130, of 19 December 2001, Technical Instruction - 13 - Emergency Lighting, Military Fire Brigade of Minas Gerais.

Minas Gerais, Law No. 14.130, of 19 December 2001, Technical Instruction - 14 - Fire Detection and Alarm System, Military Fire Brigade of Minas Gerais.

Minas Gerais, Law No. 14.130, of 19 December 2001, Technical Instruction - 15 - Emergency Signalling, Military Fire Brigade of Minas Gerais.

Minas Gerais, Law No. 14.130, of 19 December 2001, Technical Instruction - 16 - Fire Extinguisher Protection System, Military Fire Brigade of Minas Gerais.

ONO, R. Quality assurance parameters for fire safety design in high-rise buildings. Built Environment. Porto Alegre, 2007.

POZZOBON, C. E. Protection against fires and explosions: Techniques for preventing and combating accidents. Class notes. Postgraduate course in Occupational Safety Engineering. Ijuí: UNIJUI, 2007.

SEITO, Alexandre Itiu. et. al. Fire safety in Brazil. São Paulo: Projeto Editora, 2008.

CHAPTER 7

ANNEXES

7.1.CBMMG ID card

ANNEX B - FRONT - IDENTIFICATION CARD

CARTÃO DE IDENTIFICAÇÃO	Projeto N.º 001	BOMBEIRO MILITAR MINAS GERAIS
	Em 13 / 10 / 2016	O AMIGO CERTO NAS HORAS INCERTAS
	Protocolista	

Rua: Evéncio de Abreu	n.º: 34	Compl.: Edifício Axel Gomes
Bairro: Centro	Município: Matipó	UF: MG
Proprietário ou responsável p/ uso: Walquíria Schiavo Pereira	Fone:	
Técnico Responsável: Filipe Heringer Silveira do Amaral	CREA: XX	Fone: XX
Áreas - Existente: 0 m²	A construir: 453,13 m²	Total: 453,13 m²
Ocupação:		

RETIRADA DO PROJETO					
	NOTIFICAÇÃO	Em __/__/__	Nome:		RG:
			Assinatura:		Fone:
		Em __/__/__	Nome:		RG:
			Assinatura:		Fone:
		Em __/__/__	Nome:		RG:
			Assinatura:		Fone:
	APROV.	Em __/__/__	Nome:		RG:
			Assinatura:		Fone:

Aprovado em __/__/____	Analista	Ch. Seç de Análise

ANNEX B - BACK

VISTORIAS		
Protocolo nº	data __/__/__	Atendente
Vistoriador:		Parecer
Protocolo nº	data __/__/__	Atendente
Vistoriador:		Parecer
Protocolo nº	data __/__/__	Atendente
Vistoriador:		Parecer
Protocolo nº	data __/__/__	Atendente
Vistoriador:		Parecer
Protocolo nº	data __/__/__	Atendente
Vistoriador:		Parecer
Protocolo nº	data __/__/__	Atendente
Vistoriador:		Parecer

AVCB		
Protocolo nº	AVCB nº	Ch S Vistoria
Em __/__/__		Ass.:
Retirado por:		Fone:
RG:		
Protocolo nº	AVCB nº	Ch S Vistoria:
Em __/__/__		Ass.:
Retirado por:		Fone:
RG:		
Protocolo nº	AVCB nº	Ch S Vistoria:
Em __/__/__		Ass.:
Retirado por:		Fone:
RG:		

7.2. Fire and panic safety form for PTS

BOMBEIRO MILITAR MINAS GERAIS	**FORMULÁRIO DE SEGURANÇA CONTRA INCÊNDIO E PÂNICO PARA PTS**

1. IDENTIFICAÇÃO DA EDIFICAÇÃO E/OU ÁREA DE RISCO		

Logradouro Público: Rua Evêncio de Abreu

N.º: 34 Complemento: Edifício Axel Gomes	Lote: 34	Quarteirão:
Bairro: Centro CEP: 35367 - 000	Município: Matipó	UF: MG
Proprietário: Walquíria Schiavo Pereira	CPF/CNPJ:	Fone: ( )
Responsável pelo uso: Walquíria Schiavo Pereira	CPF/CNPJ :	Fone: ( )
Existente: 453,13 m² A construir: 0 m²	Total: 453,13 m²	
Altura: 10,5 m n.º de pav.: 04 Ocupação do subsolo:		
Uso, divisão e descrição: A – 2, Edificação multifamiliar	Carga Incêndio (MJ/m²): 300	

2. ELEMENTOS ESTRUTURAIS

Estrutura portante (concreto, aço, madeira, outros): Concreto e aço

Estrutura de sustentação da cobertura (concreto, aço, madeira, outros): Concreto e aço

3. FORMA DE APRESENTAÇÃO	Protocolo (uso do CBMMG)
Projeto Técnico Simplificado	

4. MEDIDAS DE SEGURANÇA CONTRA INCÊNDIO E PÂNICO			
Controle de materiais de acabamento	X	Sinalização de emergência	
Saídas de emergência	X	Extintores	
Iluminação de emergência	X		

5. RISCOS ESPECIAIS			
Armazenamento de líquidos inflamáveis/combustíveis		Fogos de artifício	
Gás Liqüefeito de Petróleo		Vaso sob pressão (caldeira)	
Armazenamento de produtos perigosos		Outros (especificar)	

______________________________ Ass: Proprietário ou Responsável pelo uso	______________________________ Ass: Vistoriador do Corpo de Bombeiros
______________________________ Ass: Responsável Técnico	______________________________ Ass:Chefe da Seção de Vistoria

7.3.
Technical projects

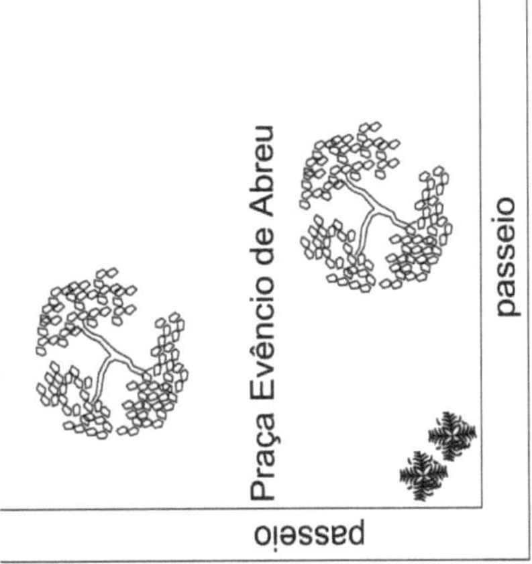

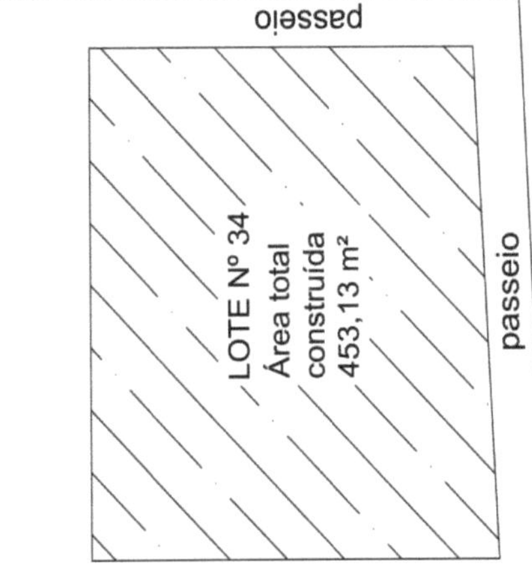

passeio

SITUAÇÃO
Área do terreno 98,76 m²
Área total construída 453,13 m²
Sem escala

39

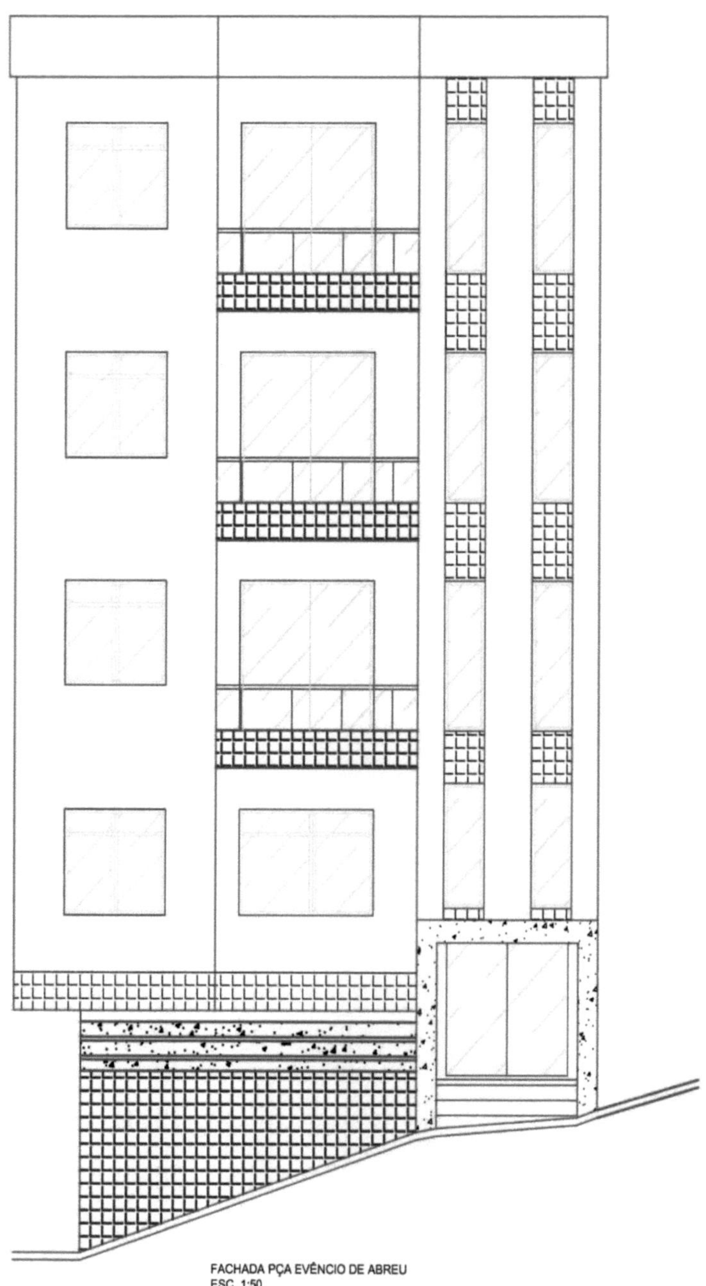

FACHADA PÇA EVÊNCIO DE ABREU
ESC. 1:50

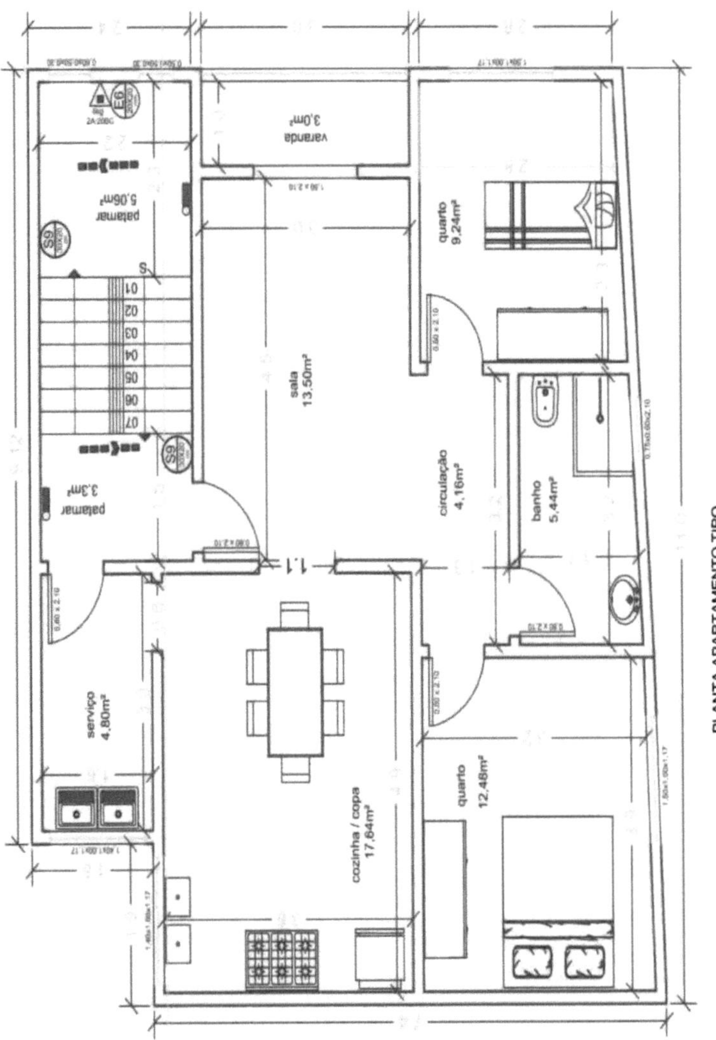

PLANTA APARTAMENTO TIPO
ESC. 1:50
ÁREA CONSTRUÍDA 89,50 m²

41

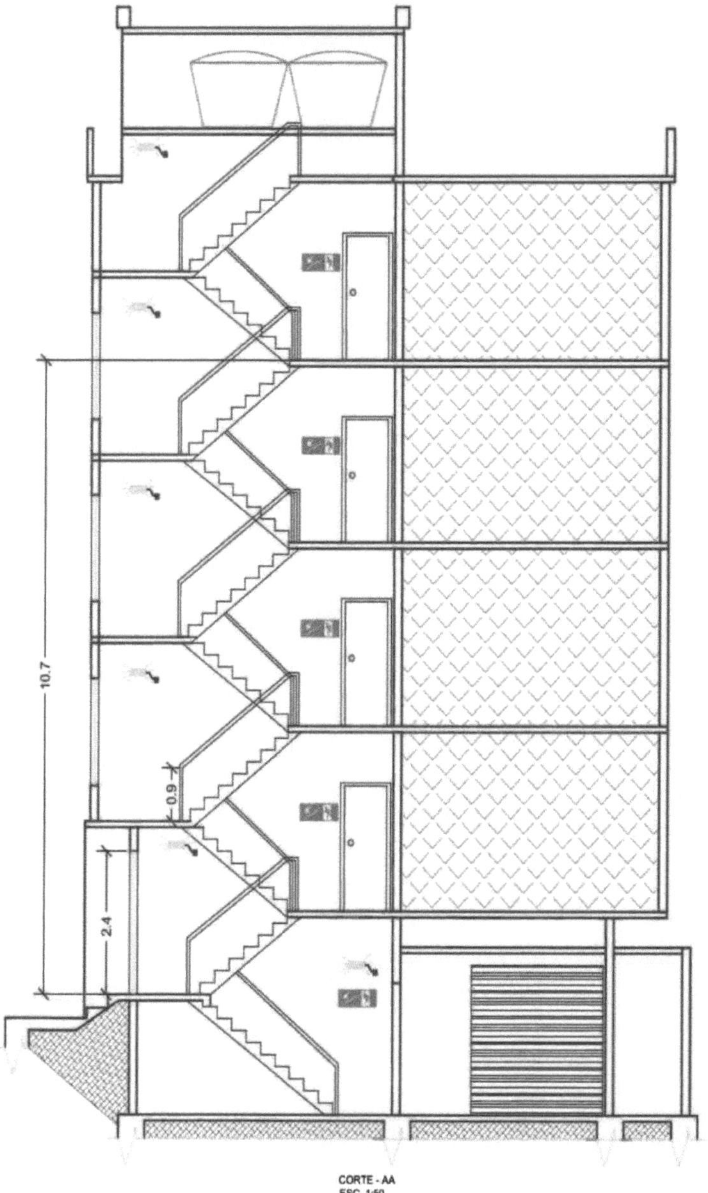
10.7
0.9
2.4
CORTE - AA
ESC. 1:50

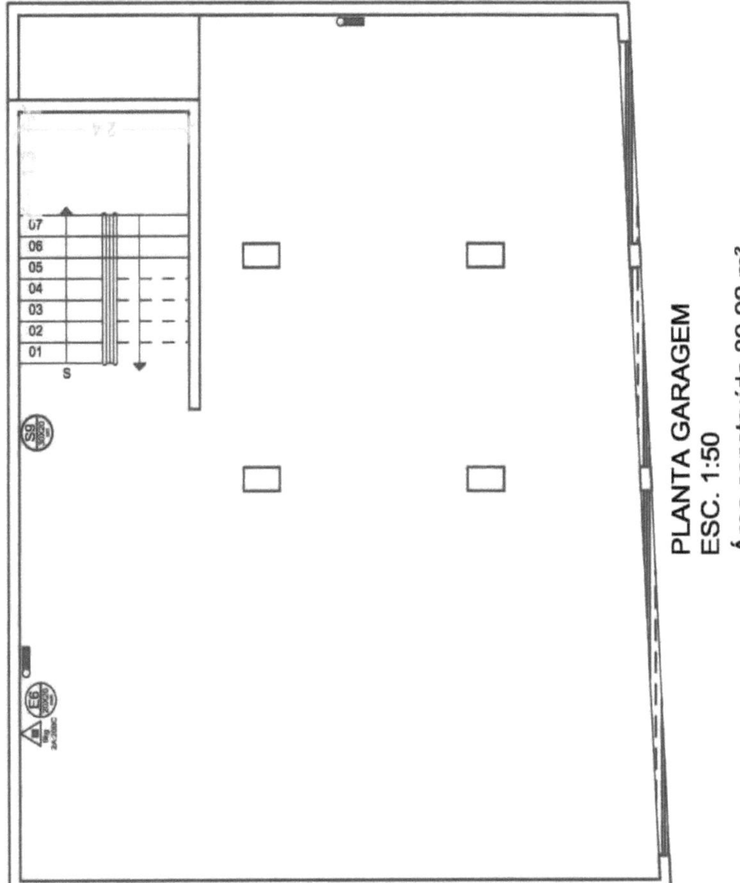
07
06
05
04
03
02
01
S
PLANTA GARAGEM
ESC. 1:50
Área construída 89,92 m²

SAFETY MEASURES TAKEN FOR THIS BUILDING

EXTINTORES	CONFORME IT 18 - SISTEMA DE PROTEÇÃO POR EXTINTORES. UTILIZAM EXTINTORES - PÓ ABC - 2A:20BC:6KG
ILUMINAÇÃO DE EMERGÊNCIA	CONFORME IT - 18 - ILUMINAÇÃO DE EMERGÊNCIA. UTILIZAR LÂMPADAS DE EMERGÊNCIA COM ACENDIMENTO AUTOMÁTICO h 250cm
SINALIZAÇÃO DE EMERGÊNCIA	CONFORME IT 15 - SINALIZAÇÃO DE ROTA DE FUGA E SINALIZAÇÃO DE SAÍDA DE EMERGÊNCIA - UTILIZADA PLACAS FOTOLUMINESCENTES NO SENTIDO DE FUGA
SAÍDAS DE EMERGÊNCIA	CONFORME IT 08 - SAÍDA DE EMERGÊNCIA NAS EDIFICAÇÕES. AS SAÍDAS DE EMERGÊNCIA ATENDEM AS NORMAS VIGENTES.

CLASSIFICAÇÃO - NBR 9077/2001

GRUPO	OCUPAÇÃO	DIVISÃO	DESCRIÇÃO	EXEMPLOS
A	EDIFICAÇÃO MULTIFAMILIAR	A-2	PRÉDIO RESIDENCIAL	APARTAMENTO

CARGA DE INCÊNDIO - IT 09

OCUPAÇÃO/USO	DESCRIÇÃO	DIVISÃO	CARGA DE INCÊNDIO EM MJ/m²
A EDIFICAÇÃO MULTIFAMILIAR	PRÉDIO RESIDENCIAL	A - 2	300 MJ/m²

CLASSIFICAÇÃO DAS EDIFICAÇÕES E ÁREAS DE RISCO QUANTO A CARGA DE INCÊNDIO

RISCO	CARGA DE INCÊNDIO EM MJ/m²
BAIXO	300 MJ/m²

FACULDADE VÉRTICE - UNIVÉRTIX

SOCIEDADE EDUCACIONAL GARDINGO LTDA- SOEGAR

THESIS PROTOCOL FOR EXAM BOARDS

To Prof. Esp. Ailton Moreira Magalhães - Coordinator of the Civil Engineering Course

To Prof Pedro Genuíno Santana Júnior - Professor of the subject

Subject: Protocol of the final version of the Civil Engineering undergraduate course at the Univértix Faculty for presentation to the board.

In accordance with the discipline's compliance rules, I admit that the work entitled, **CASE STUDY IN A BUILDING IN THE CITY OF MATIPÓ - MG: DIMENSIONING OF THE FIRE PREVENTION AND PROTECTION PLAN**, carried out by the student Filipe Heringer Silveira do Amaral, contains all the elements required in the basic rules of the course's TC document, and is in a position to be filed.

I hereby certify that this student is able to defend their research and that, as their supervisor, I authorise them to present it.

Matipó, 11th November 2016.

Yours sincerely,

____

Prof Pedro Genuíno Santana Junior ORIENTING PROFESSOR

MIX
Papier aus verantwortungsvollen Quellen
Paper from responsible sources
FSC® C105338

Printed by Books on Demand GmbH, Norderstedt / Germany